An Introduction to Multiplication

Frank Romascavage, III, PhD

fromascavage@yahoo.com

Cover design by Frank Romascavage, III

Paperback ISBN: 978-1-951816-04-9

Ebook ISBN: 978-1-951816-05-6

Introduction

The multiplication expressions from 0×0 to 9×9 as well as their respective products are presented in the first part of this text. On the front of the page you will be given a multiplication expression, for example 2×3. Next, you will have to determine the product of the expression on that page. Then, you will check to see if your answer is correct by flipping to the back of the page. On the back of the page, you will find the correct answer. For example, 6 is on the back of the page that has 2×3 on the front, because 2×3 is equal to 6. If you are reading this text electronically, the answer to the multiplication expression is on the following page. For example, 6 is on the page that follows 2×3.

This text is recommended for anyone who wants to improve their proficiency in multiplying single digits. You do not have to start at 0×0 and proceed in order to 9×9 to get the most out of this text. Feel free to skip to any expressions that you want to study. For example, if you want to review 5×7, you do not have to solve all of the problems from 0×0 to 5×6 to get there. You can proceed directly to 5×7.

Ten worksheets are provided after all of the multiplication expressions and their products are presented. Each worksheet has ten questions consisting solely of multiplying single digits together. By doing the worksheets, you will have the opportunity to assess what you have learned through additional practice. Last, solutions to the worksheets will be provided at the end of the text. I hope this text will greatly increase your abilities and confidence while multiplying single digits!

0×0

0

0×1

0

0×2

0

0×3

0

0×4

0

0×5

0

0×6

0

$$0 \times 7$$

0

0×8

0

0 × 9

0

1 × 0

0

1 × 1

1

1 × 2

2

1 × 3

3

1×4

4

1 × 5

5

1×6

6

1 × 7

7

1 × 8

8

1×9

9

2×0

0

2 × 1

2

2×2

4

$$2 \times 3$$

6

2 × 4

8

2×5

10

$$2 \times 6$$

12

2 × 7

14

$$2 \times 8$$

16

$$2 \times 9$$

18

$$3 \times 0$$

0

3 × 1

3

$$3 \times 2$$

6

3×3

9

3×4

12

$$3 \times 5$$

15

3 × 6

18

$$3 \times 7$$

21

3×8

24

$$3 \times 9$$

27

4×0

0

4 × 1

4

$$4 \times 2$$

8

4 × 3

12

4×4

16

4×5

20

4 × 6

24

$$4 \times 7$$

28

4 × 8

32

$$4 \times 9$$

36

$$5 \times 0$$

0

5 × 1

5

5×2

10

5×3

15

5 × 4

20

5 × 5

25

$$5 \times 6$$

30

5 × 7

35

5 × 8

40

5×9

45

6×0

0

6 × 1

6

$$6 \times 2$$

12

6×3

18

6×4

24

$$6 \times 5$$

30

6×6

36

$$6 \times 7$$

42

6×8

48

6 × 9

54

$$7 \times 0$$

0

$$7 \times 1$$

7

7×2

14

7×3

21

7×4

28

7×5

35

$$7 \times 6$$

42

7 × 7

49

7×8

56

$$7 \times 9$$

63

8×0

0

8×1

8

8×2

16

8×3

24

8×4

32

8×5

40

8 × 6

48

$$8 \times 7$$

56

8 × 8

64

8 × 9

72

$$9 \times 0$$

0

9×1

9

9×2

18

$$9 \times 3$$

27

9×4

36

9×5

45

$$9 \times 6$$

54

$$9 \times 7$$

63

$$9 \times 8$$

72

$$9 \times 9$$

81

Worksheet #1

$4 \times 7 =$

$1 \times 1 =$

$3 \times 0 =$

$9 \times 2 =$

$5 \times 6 =$

$2 \times 2 =$

$8 \times 3 =$

$7 \times 7 =$

$0 \times 4 =$

$6 \times 5 =$

Worksheet #2

$3 \times 6 =$

$7 \times 9 =$

$1 \times 2 =$

$0 \times 8 =$

$5 \times 3 =$

$4 \times 5 =$

$8 \times 6 =$

$9 \times 9 =$

$6 \times 0 =$

$2 \times 7 =$

Worksheet #3

$7 \times 1 =$

$0 \times 5 =$

$2 \times 4 =$

$3 \times 3 =$

$5 \times 8 =$

$9 \times 7 =$

$6 \times 6 =$

$4 \times 1 =$

$8 \times 2 =$

$1 \times 9 =$

Worksheet #4

$4 \times 0 =$

$7 \times 5 =$

$9 \times 8 =$

$3 \times 1 =$

$6 \times 7 =$

$5 \times 4 =$

$2 \times 3 =$

$1 \times 8 =$

$0 \times 0 =$

$8 \times 7 =$

Worksheet #5

$1 \times 3 =$

$2 \times 6 =$

$3 \times 4 =$

$7 \times 2 =$

$5 \times 9 =$

$8 \times 0 =$

$4 \times 4 =$

$0 \times 1 =$

$6 \times 9 =$

$9 \times 6 =$

Worksheet #6

$2 \times 8 =$

$0 \times 3 =$

$4 \times 9 =$

$7 \times 8 =$

$8 \times 1 =$

$1 \times 5 =$

$9 \times 3 =$

$6 \times 8 =$

$3 \times 7 =$

$5 \times 5 =$

Worksheet #7

$1 \times 4 =$

$0 \times 6 =$

$2 \times 9 =$

$8 \times 5 =$

$7 \times 4 =$

$3 \times 2 =$

$4 \times 6 =$

$5 \times 2 =$

$9 \times 0 =$

$6 \times 1 =$

Worksheet #8

$8 \times 8 =$

$0 \times 7 =$

$6 \times 4 =$

$7 \times 3 =$

$9 \times 5 =$

$5 \times 1 =$

$4 \times 2 =$

$3 \times 9 =$

$2 \times 1 =$

$1 \times 0 =$

Worksheet #9

$7 \times 0 =$

$4 \times 8 =$

$6 \times 3 =$

$5 \times 7 =$

$8 \times 9 =$

$0 \times 2 =$

$3 \times 5 =$

$1 \times 6 =$

$2 \times 5 =$

$9 \times 4 =$

Worksheet #10

$0 \times 9 =$

$1 \times 7 =$

$2 \times 0 =$

$3 \times 8 =$

$4 \times 3 =$

$5 \times 0 =$

$6 \times 2 =$

$7 \times 6 =$

$8 \times 4 =$

$9 \times 1 =$

Answers to Worksheet #1

$4 \times 7 = 28$

$1 \times 1 = 1$

$3 \times 0 = 0$

$9 \times 2 = 18$

$5 \times 6 = 30$

$2 \times 2 = 4$

$8 \times 3 = 24$

$7 \times 7 = 49$

$0 \times 4 = 0$

$6 \times 5 = 30$

Answers to Worksheet #2

$3 \times 6 = 18$

$7 \times 9 = 63$

$1 \times 2 = 2$

$0 \times 8 = 0$

$5 \times 3 = 15$

$4 \times 5 = 20$

$8 \times 6 = 48$

$9 \times 9 = 81$

$6 \times 0 = 0$

$2 \times 7 = 14$

Answers to Worksheet #3

$7 \times 1 = 7$

$0 \times 5 = 0$

$2 \times 4 = 8$

$3 \times 3 = 9$

$5 \times 8 = 40$

$9 \times 7 = 63$

$6 \times 6 = 36$

$4 \times 1 = 4$

$8 \times 2 = 16$

$1 \times 9 = 9$

Answers to Worksheet #4

$4 \times 0 = 0$

$7 \times 5 = 35$

$9 \times 8 = 72$

$3 \times 1 = 3$

$6 \times 7 = 42$

$5 \times 4 = 20$

$2 \times 3 = 6$

$1 \times 8 = 8$

$0 \times 0 = 0$

$8 \times 7 = 56$

Answers to Worksheet #5

$1 \times 3 = 3$

$2 \times 6 = 12$

$3 \times 4 = 12$

$7 \times 2 = 14$

$5 \times 9 = 45$

$8 \times 0 = 0$

$4 \times 4 = 16$

$0 \times 1 = 0$

$6 \times 9 = 54$

$9 \times 6 = 54$

Answers to Worksheet #6

$2 \times 8 = 16$

$0 \times 3 = 0$

$4 \times 9 = 36$

$7 \times 8 = 56$

$8 \times 1 = 8$

$1 \times 5 = 5$

$9 \times 3 = 27$

$6 \times 8 = 48$

$3 \times 7 = 21$

$5 \times 5 = 25$

Answers to Worksheet #7

$1 \times 4 = 4$

$0 \times 6 = 0$

$2 \times 9 = 18$

$8 \times 5 = 40$

$7 \times 4 = 28$

$3 \times 2 = 6$

$4 \times 6 = 24$

$5 \times 2 = 10$

$9 \times 0 = 0$

$6 \times 1 = 6$

Answers to Worksheet #8

$8 \times 8 = 64$

$0 \times 7 = 0$

$6 \times 4 = 24$

$7 \times 3 = 21$

$9 \times 5 = 45$

$5 \times 1 = 5$

$4 \times 2 = 8$

$3 \times 9 = 27$

$2 \times 1 = 2$

$1 \times 0 = 0$

Answers to Worksheet #9

$7 \times 0 = 0$

$4 \times 8 = 32$

$6 \times 3 = 18$

$5 \times 7 = 35$

$8 \times 9 = 72$

$0 \times 2 = 0$

$3 \times 5 = 15$

$1 \times 6 = 6$

$2 \times 5 = 10$

$9 \times 4 = 36$

Answers to Worksheet #10

$0 \times 9 = 0$

$1 \times 7 = 7$

$2 \times 0 = 0$

$3 \times 8 = 24$

$4 \times 3 = 12$

$5 \times 0 = 0$

$6 \times 2 = 12$

$7 \times 6 = 42$

$8 \times 4 = 32$

$9 \times 1 = 9$

Made in the USA
Monee, IL
07 November 2026

68010596R00136